Astronom

CONTENTS

Alpha Omega Publications®

804 N. 2nd Ave. E., Rock Rapids, IA 51246-1759
© MCMXCVII by Alpha Omega Publications, Inc. All rights reserved.
LIFEPAC is a registered trademark of Alpha Omega Publications, Inc.

Dear Instructor,

Thank you for your interest in electives using the LIFEPAC Select Series.

The courses in this series have been compiled by schools using Alpha Omega's LIFEPAC Curriculum. These courses are an excellent example of the flexibility of the LIFEPAC Curriculum for specialized teaching purposes.

The unique design of the worktext format has allowed instructors to mix and match LIFEPACs from four subjects (Bible, Language Arts, Science, and History & Geography) to create alternative courses for junior high and high school credit.

These courses work particularly well as unit studies, as supplementary electives, or for meeting various school and state requirements. Another benefit of the courses—and any LIFEPAC subject, for that matter—is the ability to use them with any curriculum, at any time during the year, for any of several purposes:

- Elective Courses
- Make-up Courses
- Substitution Courses
- Unit Studies

- Summer School Courses
- Remedial Courses
- Multi-level Teaching
- Thematic Studies

Course Titles	Suggested Credits
Astronomy (Jr. High and above)	$\frac{1}{2}$ credit
Composition	$\frac{1}{2}$ credit
Geography	$\frac{1}{2}$ credit
Geology	$\frac{1}{2}$ credit
Life of Christ (Jr. High and above)	$\frac{1}{2}$ credit
Life Science	$\frac{1}{2}$ credit
Mankind: Anthropology and Sociology	$\frac{1}{2}$ credit

Astronomy

Jr. High Level and above (1/2 credit)

Spaceship Earth
Science LIFEPAC 608

Earth's Motion
- Earth's Size & Shape
- Earth's Rotation
- Earth's Revolution
- Time

Eclipses
- Solar Eclipse
- Lunar Eclipse

Our Solar System
- The Sun
- Planets and Light-Years
- Asteroids, Comets, and Meteors

Astronomy and The Stars
Science LIFEPAC 609

Astronomy
- God, Astronomy, & the Universe
- History
- Astronomy Today

Stars
- Colors and Temperatures
- Giants and Dwarfs
- Elements and Spectra
- Magnitude and Luminosity
- Light-Years and Astronomical units

Constellations and Major Stars
- Familiar Constellations
- Major Stars

Earth In Space 1
Science LIFEPAC 703

Stargazing
- The Ancients
- Constellations
- Geocentric Theory
- Wanderers
- Meteors

Astronomy
- The Transparent Sphere Hypothesis
- Copernicus
- The Telescope
- Galileo Galilei
- Tycho Brahe
- Johannes Kepler
- A Change of Mind
- Isaac Newton
- Measuring the Sky
- The Astronomer
- The Astronomer's Tools

Earth In Space 2
Science LIFEPAC 704

The Sun's Energy
- Benefits of Solar Energy
- Source of Solar Energy

The Sun's Family
- Inner Planets
- Outer Planets
- Asteroids
- Comets

The Moon
- Orbit
- Rise and Set
- Rotation
- Phases
- Tides

Eclipses
- Why Eclipses Occur
- When Eclipses Occur
- What Kinds of Eclipses Occur

Astronomy
Science LIFEPAC 907

The Universe
- Extent of the Universe
- Constellations

Telescopes and Optics
- Magnification
- Types of Telescopes
- Newer Types of Telescopes

Space Explorations
- Requirements for Launching a Satellite
- Planetary Missions

Materials Needed for LIFEPAC

Required:

coin, such as a quarter or a half dollar
a large ball about the size of a basketball *or* a cardboard circle 8 inches diameter
a small ball about the size of a tennis ball *or* a cardboard circle 3 inches diameter
a light of about 100 watts or more
yardstick
ruler
piece of chalk
ball of string or twine
150 pennies

Additional Learning Activities

Section I Earth's Motion

1. With a friend or an adult, use a globe and light bulb to demonstrate how the sun shines on the earth.

2. With a friend, demonstrate how the angle of the sun's rays affects the amount of heat the earth gets. Fill two boxes with sand or dirt. Lay a thermometer in each box with its bulb buried about an inch deep. Set one box flat on the ground and the other on a slant. Check the temperature in each box in about ten minutes.

3. In your own words, explain why an accident could happen at 8:00 A.M. and someone in Washington, D.C., could hear about it before 4:00 A.M.

4. Write a story about what it would be like to live on a part of the earth where the sun doesn't set for months and then doesn't rise for months.

Section II Eclipses

1. Arrange a field trip to go to a local planetarium.

2. With a friend, make two posters. On one poster, make a drawing of a lunar eclipse. On the other, make a drawing of a solar eclipse.

3. Do some research on the solar eclipse. When was the last one where you live? See how many photographs you can find. Display them.

Section III The Solar System

1. With a friend, make a model of the solar system using different sized balls. Label all the planets and the sun.

2. With a friend, make a large chart that shows a comparison of the planets. Include information like diameter, distance from the sun, number of satellites, and revolution and rotation times. Display your chart in your classroom.

Independent Activities

1. Select one planet (other than earth), and write a two-page report on it.

2. Read science fiction stories about life on other planets. Could such life really exist? Explain your reason in a one-page report.

SECTION ONE

1.1   a. 7,900 miles

       b. 7,926 miles

       c. 26 miles

       d. no

1.2   true

1.3   false

1.4   true

1.5   true

1.6   true

1.7   false

1.8   false

1.9   true

1.10   The rotation of the earth about its axis causes night and day.

1.11   In relation to the sun, the earth takes 24 hours to make one complete rotation about its axis. This is called a *solar day*. In relation to the very same position on earth compared to far distant stars, however, the earth takes 23 hours 56 minutes 4.091 seconds to make one complete rotation. This is called a *sidereal day*.

1.12   The distance around the earth parallel to the equator varies, depending on where you are located. At the equator, the distance around the earth is the greatest. As you move toward either pole, the distance is smaller. Therefore, you travel less in a day as you move toward the poles from the equator, and the speed of motion would be less.

1.13   a. sunset

1.14   c. midnight

1.15   d. 1800

1.16   d. 360

1.17   b. Standard time

1.18   b. Prime Meridian

1.19   a. International Date Line

1.20   b. four

1.21   f

1.22   h

1.23   g

1.24   a

1.25   j

1.26   b

1.27   k

1.28   c

1.29   l

1.30   d

1.31   The day and night are an equal amount of time. Also, the sun is directly overhead at noon on the equator.

1.32   In the Northern hemisphere, after the vernal equinox, the days become longer. After the autumnal equinox, the days become shorter.

1.33   No. Scientific study is limited by the facts that mankind has observed on the earth, in the lab, and in his explorations with a microscope or telescope. Mankind can only imagine what conditions exist on a planet orbiting a star in a faraway solar system or galaxy.

1.34   Adult check

SECTION TWO

2.1 There is less of the tree visible, until it is eventually blocked from sight.

2.2 a. moon
Either order:
b. earth
c. sun

2.3 a. two
b. three

2.4 They have been able to determine the exact relative positions of the earth, sun, and moon. They have been able to study possible changes in the strength of gravity and the size of the sun. The size of distant stars has been determined through the study of eclipses of other heavenly bodies.

2.5 true

2.6 false

2.7 true

2.8 true

2.9 false

2.10 true

2.11 false

2.12 b

2.13 a

2.14 c

2.15 b

2.16 c

2.17 a

2.18 the earth, moon, and sun are nearly in a straight line and the moon passes between the earth and sun.

2.19 a total or partial obscuring (or darkening) of one celestial body by another.

2.20 the earth, moon, and sun are nearly in a straight line and the moon passes through the earth's shadow.

2.21 pertaining to the moon.

2.22 pertaining to the sun.

2.23 During a solar eclipse, the line-up is sun, moon, and earth. During a lunar eclipse, the line-up is sun, earth, and moon.

2.24 a. opinion
b. fact
c. fact
d. fact
e. opinion
f. opinion
g. opinion
h. fact
i. opinion
j. fact
k. fact
l. fact

2.25 a. syl lab i ca tion
b. rev o lu tion
c. sat el lite
d. ro ta tion
e. ev o lu tion
f. hor i zon tal
g. ver ti cal ly
h. el lip ti cal ly
i. grav i ta tion
j. in er tia

SECTION THREE

3.1–3.3 Answers may vary. The answers given are close approximations.

3.1 109

3.2 863,934 miles

3.3 a. 863,964
     b. 865,000

3.4 star

3.5 99

3.6 Either order:
     a. hydrogen
     b. helium

3.7 nuclear fusion

3.8 10,000° F

3.9 solar activity

3.10 cooler

3.11 prominences

3.12 Solar flares

3.13 northern and southern lights

3.14 true

3.15 false

3.16 true

3.17 true

3.18 true

3.19 Teacher/Adult check (Answers will vary based on rounding and the source of the information.)

3.19

Standard Chart

Planets	Diameter (miles)	Distance from the sun (mean)	Number of Moons	Length of Day (hrs)	Length of Year (yrs)
Mercury	3,031	36,000,000	0	1,416	.24
Venus	7,520	67,000,000	0	5,832	.6
Earth	7,926	93,000,000	1	24	1
Mars	4,220	141,700,000	2	24.7	1.8
Jupiter	88,700	483,700,000	67	8.9	12
Saturn	74,600	885,200,000	62	10.7	29.5
Uranus	31,570	1,781,000,000	27	17.2	84
Neptune	30,200	2,788,000,000	14	16.1	165

3.19

Metric Chart

Planet	Diameter (km)	Distance (10^6 km)	Moons	Length of Day (hrs)	Length of Year (days)
Mercury	4,879	57.9	0	1,416	88
Venus	12,104	108.2	0	5,832	224.7
Earth	12,756	149.6	1	24	365.2
Mars	6,792	227.9	2	24.7	687
Jupiter	142,984	778.6	67	9.9	4,331
Saturn	120,536	1,433.5	62	10.7	10,747
Uranus	51,118	2,872.5	27	17.2	30,589
Neptune	49,528	4,495.1	14	16.1	59,800

3.20   Adult check

3.21   a. Mercury
       b. Venus
       c. Earth
       d. Mars
       e. Jupiter
       f. Saturn
       g. Uranus
       h. Neptune

3.22   h
3.23   i
3.24   f
3.25   g
3.26   a
3.27   e
3.28   b

SELF TEST 1

1.01   l

1.02   k

1.03   j

1.04   a

1.05   b

1.06   c

1.07   d

1.08   i

1.09   h

1.010   f

1.011   true

1.012   true

1.013   true

1.014   false

1.015   false

1.016   true

1.017   true

1.018   true

1.019   false

1.020   false

1.021   b. $23\frac{1}{2}°$

1.022   a. sunset

1.023   a. 15°

1.024   c. Greenwich, England

1.025   c. 5:00 A.M.

1.026   c. 366

1.027   c. winter

1.028   a. ellipse or oval

1.029   c. reversed

1.030   c. the winter

1.031   c. 101,000

1.032   Any order:

   a. rotates about its axis

   b. orbits around the sun

   c. moves with sun around the center of the Milky Way Galaxy

   d. moves with Milky Way Galaxy through the universe

1.033   Night and day occur as the earth rotates about it axis. The sun shines on one side of the earth as it rotates (daylight), and the other side of earth is hidden from the sun (night).

1.034   The four seasons are spring, fall, summer, and winter. The seasons occur because the earth is tilted on its axis at $23\frac{1}{2}°$ and the earth is orbiting around the sun. When the Northern Hemisphere is tilted more toward the sun, it is summer in the Northern Hemisphere. The temperatures are then warmer. In the winter, the Northern Hemisphere is tilted away from the sun. The temperatures are then colder.

1.035   At the vernal equinox, the sun is directly over the equator at noon, and the day and night are equal length. As the earth continues to orbit around the sun, the days get longer in the Northern Hemisphere. At the autumnal equinox, the sun is directly overhead the equator at noon and the days and nights are equal length. As the earth continues to orbit around the sun, the days in the Northern Hemisphere get shorter.

SELF TEST 2

2.01   d

2.02   h

2.03   e

2.04   a

2.05   g

2.06   c

2.07   f

2.08   b

2.09   false

2.010   true

2.011   false

2.012   false

2.013   false

2.014   true

2.015   true

2.016   false

2.017   Either answer:
        inertia
        gravity

2.018   93,000,000 miles

2.019   Any order:
        a. Eastern
        b. Central
        c. Mountain
        d. Pacific

2.020   23½°

2.021   leap year

2.022   the earth, moon, and sun are nearly in a straight line and the moon passes between the earth and sun.

2.023   the earth, moon, and sun are nearly in a straight line and the moon passes through the earth's shadow.

2.024   orbit

2.025   a. the earth

2.026   c. solar eclipse

2.027   c. summer

2.028   b. 595 million

2.029   c. vernal

2.030   During a solar eclipse, the line-up is sun, moon, and earth. During a lunar eclipse, the line-up is sun, earth, and moon.

2.031   A solar eclipse is rare because the moon's orbit is tilted about five degrees away from the earth's orbit. The earth's shadow and the moon's shadow usually go into space without touching either body. The moon and the earth travel either above or below, not in, the shadow of the other.

SELF TEST 3

3.01 false

3.02 true

3.03 false

3.04 true

3.05 true

3.06 true

3.07 false

3.08 false

3.09 false

3.010 false

3.011 true

3.012 g

3.013 f

3.014 l

3.015 i

3.016 h

3.017 e

3.018 k

3.019 j

3.020 c

3.021 b

3.022 a. Mercury
b. Venus
c. Earth
d. Mars
e. Jupiter
f. Saturn
g. Uranus
h. Neptune

3.023 c. 99

3.024 b. nuclear fusion

3.025 a. 10,000

3.026 b. magnetic fields

3.027 b. 25 days 9 hours

3.028 b. prominences

3.029 a. Northern Lights

3.030 a. no

3.031 b. 200 or more years

3.032 c. 500

3.033 Comets are bright objects in space with a nucleus made up of water, carbon dioxide, methane, and ammonia, with dust and rocky material buried in them. They have long, elliptical orbits around the sun. Gas around the nucleus given off is called the coma. Dust and gas given off that streams into space is called the tail.

3.034 A meteoroid consists of particles of dust and very small rocks in space. When they enter the earth's atmosphere, they are called meteors or "shooting stars." When these hit the earth, they are called meteorites.

1. j

2. k

3. a

4. b

5. m

6. l

7. c

8. i

9. d

10. h

11. e

12. false

13. false

14. true

15. true

16 true

17. true

18. true

19. false

20. true

21. false

22. false

23. a. Mercury
   b. Venus
   c. Earth
   d. Mars
   e. Jupiter
   f. Saturn
   g. Uranus
   h. Neptune

24. Any order:
   a. rotates about its axis
   b. orbits around the sun
   c. moves with sun around the center of the Milky Way Galaxy
   d. moves with Milky Way Galaxy through the universe

25. b. 23½°

26. a. sunset

27. 101,000

28. a. the earth

29. c. solar eclipse

30. c. 99

31. a. helium

32. b. 25 days 9 hours

33. From the rising of the sun unto the going down of the same the Lord's name is to be praised.

34. Night and day occur as the earth rotates about it axis. The sun shines on one side of the earth as it rotates (daylight), and the other side of earth is hidden from the sun (night).

35. The four seasons are spring, fall, summer, and winter. The seasons occur because the earth is tilted on its axis at $23\frac{1}{2}°$ and the earth is orbiting around the sun. When the Northern Hemisphere is tilted more toward the sun, it is summer in the Northern Hemisphere. The temperatures are then warmer. In the winter, the Northern Hemisphere is tilted away from the sun. The temperatures are then colder.

Name _____

Match each item (each answer, 2 points).

1. ____ an eclipse in which the moon is darkened   a. an asteroid

2. ____ an eclipse in which the sun is darkened   b. nuclear fusion

3. ____ the lightest, outer edges of a shadow   c. Earth

4. ____ a law that pertains to heavenly bodies being attracted to one another   d. lunar eclipse

5. ____ a shooting star   e. solar eclipse

6. ____ a planet with rings   f. meteor

7. ____ the most distant planet   g. Neptune

8. ____ a planet with no moons   h. Venus

9. ____ the star closest to Earth   i. penumbra

10. ____ the sun's energy   j. gravitation

11. ____ the water planet   k. Saturn

     l. the sun

Write *true* **or** *false* (each answer, 1 point).

12. _____ The earth is about 93,000,000 miles from the sun.

13. _____ The whole earth is darkened during a total eclipse

14. _____ Most asteroids are located between Mars and Jupiter.

15. _____ A solar day is longer than a sidereal day.

16. _____ Jupiter is the largest planet in our solar system.

17. _____ The South Pole has six months of darkness.

Complete these statements (each answer, 3 points).

18. Days and nights are equal during the _____ .

19. A *shooting star* that hits the earth is called a _____ .

20. The diameter of the earth at the poles is about _____ miles.

21. Earth is a very special place because it has a. _____ and b. _____ .

22. Our solar system contains _____ planets.

23. People within the penumbra see a _____ eclipse of the sun.

24. The earth is tilted on its axis about _____ .

25. During a _____ eclipse, the earth is darkened because the moon comes between the earth and the sun.

26. The sun appears to rise in the east and set in the west because of _____
_____
_____ .

27. If the earth were not tilted, the days and nights would be _____ .

28. The two laws concerning gravitation and inertia were proposed by
_____ .

29. The seasons are caused by the earth's tilt and its _____ .

30. Objects in our solar system that have a coma and tail are called _____ .

31. A _____ is a dust or small particle in space that becomes a
shooting star when it enters earth's atmosphere.

32. The temperature on the surface of the sun is about _____ .

Complete these activities (each answer, 3 points).

33. What causes a total eclipse of the sun?
_____
_____

34. Three types of motion of spaceship earth are:

    a. _____
    b. _____
    c. _____

35. How are the vernal equinox and the autumnal equinox the same?
_____
_____

36. Name the first three planets outward from the sun in their proper order.

    a. _____
    b. _____
    c. _____

80 / 100

Date _____

Score _____

1. d
2. e
3. i
4. j
5. f
6. k
7. g
8. h
9. l
10. b
11. c

12. true
13. false
14. true
15. true
16. true
17. true

18. equinox
19. meteorite
20. 7,900
21. Either order:
    a. oxygen (or air)
    b. water
22. eight
23. partial

24. $23\frac{1}{2}°$
25. solar
26. Example: its rotation about its axis from west to east (or counterclockwise motion as seen from above the north pole).
27. equal
28. Sir Isaac Newton
29. orbit around the sun
30. comets
31. meteoroid
32. 10,000 °F
33. The moon crosses the path between the earth and the sun, casting a dark shadow on the earth called the umbra.
34. Any three of these, and any order:
    a. rotates about its axis
    b. orbits around the sun
    c. moves with sun around the center of the Milky Way Galaxy
    d. moves with Milky Way Galaxy through the universe
35. The sun is directly overhead at noon at the equator. The days and nights are equal length.
36. a. Mercury
    b. Venus
    c. Earth

Materials Needed for LIFEPAC

Required:

white construction paper/poster board
coloring crayons, colored pencils, or colored markers
cardboard cylinder from inside a roll of paper towels
piece of plastic diffraction grating (*transmission type*)
small ruler
sheet of black construction paper
scotch tape or masking tape
pencil
scissors
string
plastic protractor
metal nut
world map

Suggested:

Star chart for current month and latitude location

Additional Learning Activities

Section I Astronomy

1. Visit an observatory with a friend.

2. Build your own telescope. Kits are available from scientific supply houses.

Section II Stars

1. Organize an astronomy club with your friends. Set up several star-watching sessions.

2. With a friend, make a collage space vehicle. Investigate how astronomers measure distance. Report your finds to your class.

3. Obtain a star chart for the current month and for your latitude position from the Internet or from astronomy magazines. Share them with your classmates.

Section III Constellations and Major Stars

1. With several friends, write a skit about what life would be like on another star. What would our sun look like? Present the skit in class.

2. With a friend, make several large posters of the constellations described in your LIFEPAC. Display them in your classroom.

3. Select a constellation that you have not studied yet. Research it on the Internet or library. Write a one to two-page report on what you learned. Turn in your report to your teacher.

SECTION ONE

1.1   true

1.2   false

1.3   true

1.4   true

1.5   false

1.6   true

1.7   false

1.8   false

1.9   i

1.10   j

1.11   m

1.12   l

1.13   k

1.14   b

1.15   a

1.16   c

1.17   n

1.18   h

1.19   e

1.20   g

1.21   Adult check

1.22   space travel

1.23   Hubble Space Telescope (HST)

1.24   Chandra X-ray Observatory

1.25   Very Large Array (VLA)

1.26   amateur astronomers

1.27   Adult check

SECTION TWO

2.1   Adult check

2.2   Possibly 2 or 3 red stars; depends on location, time of year, and weather. Adult check.

2.3   b

2.4   c

2.5   b

2.6   c

2.7   b

2.8   The color of a star tells approximately how hot it is. Stars range from red to blue-white in color and surface temperatures from 5,500 to 55,000 degrees Fahrenheit. Our sun is a medium yellow star with a surface temperature of about 10,000 degrees Fahrenheit.

2.9   supergiant reds

2.10   white dwarfs

2.11   medium

2.12   865,000

2.13   Betelgeuse

2.14   companion

2.15   Rigel

2.16   the sun

2.17   continuous

2.18   bright-line

2.19   bright-line

2.20   c

2.21   c

2.22   a

2.23   b

2.24   b

2.25   the colors of the spectrum

2.26   the same

2.27   continuous

2.28   This order:

a. red

b. orange

c. yellow

d. green

e. blue

f. indigo

g. violet

2.29   Any order:

a. continuous

b. dark-line

c. bright-line

2.30   Answer may vary.

Example: Astronomers have discovered the continuous spectrum, brightline spectrum, and the darkline spectrum. With the spectroscope, scientists are able to tell what a star is made of. Helium, one of the gases that the sun is composed of, was discovered in the spectrum of the sun before it was discovered on earth.

2.31   tells how bright a star appears to an observer on earth

2.32   the true or real brightness of a star

2.33   Betelgeuse

2.34   Betelgeuse

2.35   f

2.36   c

2.37   d

2.38   e

2.39   b

2.40   g

2.41   a

SECTION THREE

3.1   constellation

3.2   88

3.3   1,000 – 1,500

3.4   northern

3.5   Orion, Pleiades

3.6   southernmost

3.7   latitudes

3.8   false

3.9   true

3.10 false

3.11 true

3.12 false

3.13 true

3.14 false

3.15 false

3.16 Example:

   a. The man wore a mizar.

   b. Mizar is located in the Big Dipper

3.17 Example:

   a. Mercury is a liquid metal used in thermometers.

   b. Mercury is the smallest planet in our solar system.

3.18 Example:

   a. An artificial satellite was launched in June.

   b. Our moon is a natural satellite.

3.19 Example:

   a. Sodium chloride is a binary compound.

   b. Mizar is a binary star system.

3.20 Example:

   a. The story was about Jupiter.

   b. Jupiter is the largest planet.

3.21 a. He telleth the number of the stars; he calleth them all by their names.

   b. Thou art worthy, O Lord, to receive glory and honour and power: for thou hast created all things, and for thy pleasure they are and were created.

   c. He who made the Pleiades and Orion, and turns deep darkness into the morning, and darkens the day into night, who calls for the waters of the sea, and pours them out upon the surface of the earth, the LORD is his name. (RSV)

3.22 c

3.23 d

3.24 a

3.25 b

3.26 b

3.27 b

3.28 a

3.29 Adult check

3.30 Adult check. Example: Every place on earth can be measured by latitude lines and be able to see the same things–but from a different angle.

SELF TEST 1

1.01	e
1.02	f
1.03	h
1.04	l
1.05	k
1.06	j
1.07	d
1.08	c
1.09	g
1.010	i
1.011	true
1.012	false
1.013	true
1.014	true
1.015	false
1.016	true
1.017	false
1.018	true
1.019	true
1.020	false
1.021	4
1.022	2
1.023	5
1.024	1
1.025	3

1.026 reflecting

1.027 radio waves from outer space

1.028 Hawaii

1.029 Egyptians

1.030 Pluto

1.031 Astronomy is the science of the study of the stars, planets, and other objects that make up the universe. Astrology is an occult practice that attempts to predict human affairs and events on the earth from the positions of the stars.

1.032 Answer may vary. Adult check.

1.033 Any two of the following:

– he invented the reflecting telescope

– he discovered the law of gravitation

– he discovered that visible light can be broken down into a spectrum

1.034 Einstein revolutionized our concepts about mass, energy, space, and time through his theories of relativity. His work helps astronomers understand certain aspects of the universe, such as the way stars get their energy and produce light through the transformation of mass into energy.

SELF TEST 2

2.01   j

2.02   i

2.03   h

2.04   l

2.05   f

2.06   e

2.07   d

2.08   c

2.09   a

2.010   k

2.011   true

2.012   false

2.013   false

2.014   true

2.015   true

2.016   true

2.017   false

2.018   true

2.019   true

2.020   false

2.021   the stars, planets, and other objects in the universe

2.022   Pluto

2.023   core

2.024   radio waves from outer space

2.025   Egyptians

2.026   Either order:
   a. color
   b. temperature, size, or distance

2.027   reflecting

2.028   Fraunhofer

2.029   close

2.030   gas

2.031   The color of a star indicates its surface temperature. The hotter the star, the greater its brightness and the difference of its color.

2.032   Fraunhofer lines are lines found in the dark-line spectrum. Atoms of each chemical element produce a certain set of spectral lines (Fraunhofer lines). This knowledge enabled astronomers to identify elements that make up a star by studying the spectral lines in a star's light.

2.033   Answer may vary. Adult check.

2.034   Answer may vary. Example: With our normal eyesight, unaided by a telescope, a star may appear to be bright in comparison to other stars. However, a very bright star may appear dimmer than a less luminous star just because it is farther away from our solar system.

SELF TEST 3

3.01	Ursa Major	3.023	true
3.02	asterism	3.024	true
3.03	the Big Dipper	3.025	true
3.04	any order:	3.026	b
	a. Greeks		
	b. Romans	3.027	c
3.05	sextant	3.028	a
		3.029	c
3.06	c	3.030	b
3.07	i	3.031	c
3.08	f	3.032	b
3.09	l	3.033	b
3.010	k	3.034	c
3.011	g	3.035	a
3.012	j		
3.013	d		
3.014	a		
3.015	m		
3.016	e		
3.017	h		
3.018	b		

3.019 true

3.020 false

3.021 false

3.022 true

3.036 He telleth the number of the stars; he calleth them all by their names.

3.037 An astrolabe was used to measure the stars' angles as they appeared above the northern horizon. This was useful for navigators and sailors to determine their location.

3.038 Constellations are groups of stars within a particular region of the sky. In this sense, they are real. However, the patterns that were made up by sailors, farmers, and astronomers in ancient days were imaginary. In that sense, they are not real.

1. e

2. f

3. g

4. a

5. h

6. i

7. k

8. b

9. c

10. d

11. true

12. false

13. false

14. true

15. true

16. true

17. true

18. false

19. true

20. true

21. c. reflecting

22. a. radio waves from space

23. b. Hawaii

24. c. Pluto

25. a. core

26. b. Fraunhofer

27. b. Ursa Minor

28. c. winter

29. c. Orion

30. c. Orion

31. 4

32. 1

33. 5

34. 2

35. 3

36. Astronomy is the science of the study of the stars, planets, and other objects that make up the universe. Astrology is an occult practice that attempts to predict human affairs and events on the earth from the positions of the stars.

37. Yes, Christians can accept astronomy as a science. Astronomy can help Christians explain much about the physical universe that God has created. However, for Christians, any explanation of the universe and its workings must be compatible with our belief that God created the universe and that He is in control of it.

38. The color of a star indicates its surface temperature. The hotter the star, the greater its brightness.

39. Example: In 1993, the shuttle astronauts were sent to the Hubble Space Telescope because it had a faulty mirror that had to be replaced. The shuttle astronauts changed the mirror and the telescope can see clearer than ever.

Science 609 Alternate Test

Name _____

Match each item (each answer, 2 points).

1. ____ source of sun's energy
2. ____ Great Bear constellation
3. ____ power plant of the sun
4. ____ example of red star
5. ____ example of yellow star
6. ____ southern lights
7. ____ person who discovered radio waves from outer space
8. ____ name of the dark lines in the spectrum
9. ____ inventor of telescope
10. ____ distance light travels in one year, used as a measurement for star distances
11. ____ example of a white dwarf star
12. ____ sometimes, it is the planet farthest from sun
13. ____ constellation in which Polaris is located
14. ____ third planet from the sun
15. ____ instrument used to observe a spectrum

a. sun
b. spectroscope
c. Earth
d. nuclear fusion
e. Ursa Major
f. central core
g. Betelgeuse
h. Aurora Australis
i. Jansky
j. Lippershey
k. Fraunhofer
l. one light-year
m. companion to Sirius
n. Neptune
o. Ursa Minor

Write *true* **or** *false* (each answer, 1 point).

16. _____ Galileo made the first telescope for astronomy.
17. _____ All of Galileo's observations were made from a refracting telescope.
18. _____ Fraunhofer lines are dark lines in the spectrum.
19. _____ The smallest stars are white dwarfs.
20. _____ The Great Bear constellation is known as Ursa Minor.
21. _____ Galileo observed sunspots.
22. _____ The planet Mercury is closest to the sun.
23. _____ A spectroscope is useful in studying the spectra of stars.
24. _____ The photosphere is the halo around the sun which is seen during an eclipse.
25. _____ Our solar system is located in the center of the Milky Way.
26. _____ Cassiopeia is shaped like a giant *W* or *M*.

Complete these statements (each answer, 3 points).

27. Rigel is located in the constellation _____ .

28. One way stars differ is in _____ .

29. A light-year is approximately _____ trillion miles.

30. Polaris is located in the constellation known as Ursa _____ .

31. An astronomical unit consists of _____ .

32. A diffraction grating is _____ .

33. Betelgeuse is located in the constellation _____ .

34. Einstein developed important theories of _____ .

35. The constellation known as the Herdsman is called _____ .

36. The fourth planet from the sun outward is _____ .

37. Our sun consists of 99 percent helium and _____ gases.

38. The name of a cluster of stars is _____ (you have studied two).

39. The northern and southern lights is believed to be caused by_____
_____ .

Answer these questions (each answer, 5 points).

40. If you were lost, how might you find your way home at night?
_____
_____

41. How was helium discovered?
_____
_____
_____

42. What two things has the Lord been able to do concerning the stars that man has never been able to do? (The answer lies in Psalm 147:4.)
_____
_____

43. Why did God create all things?
_____
_____

80
100

Date _____

Score _____

1. d
2. e
3. f
4. g
5. a
6. h
7. i
8. k
9. j
10. l
11. m
12. n
13. o
14. c
15. b
16. true
17. true
18. true
19. true
20. false
21. true
22. true
23. true
24. false
25. false
26. true
27. Orion

28. Any of the following:
size
color
temperature
brightness
distance

29. 6

30. Minor

31. 92,900,000 miles

32. a piece of plastic with grooves that make a spectrum like a prism

33. Orion

34. Relativity

35. Bootes

36. Mars

37. hydrogen

38. Hyades or Pleiades

39. charged particles which enter the earth's magnetic fields

40. Example: by the star Polaris

41. Example:
It was discovered by a dark line in the spectrum. It was identified on the sun before it was identified on earth.

42. Example:
The Lord can count and name the stars.

43. Example:
He created all things for His pleasure.

Materials Needed for LIFEPAC

Required:
cardboard or heavy construction paper
scissors
paper fasteners
encyclopedia or other reference books
salt cartons, the twenty-six ounce size
flashlight
nails of different sizes
books or blocks of wood to support lenses
waxed paper or tissue paper convex
lenses of different focal lengths (different
size magnifying lenses)
candle
chewing gum or modeling clay
drinking straw
rubber bands
thread
blocks of wood (13 x 8 x 1 ½ cm),
(20 x 20 x 3 cm), (20 x 3 x 3 cm)
metal washers
pieces of tin; orange juice cans or other tin
that can easily be cut
protractors, half-circle and full-circle
nails (5 cm and 4 cm)
glue

Suggested:
styrofoam ball
umbrella
magnet
small plastic bags large enough to
hold a magnet
glass jar or other glass container
medicine dropper
microscope
concave mirror like a magnifying
make-up mirror
corrugated cardboard, any heavy
cardboard
thumbtacks
fishline or string

Additional Learning Activities

Section I Stargazing

1. With poster board and paints or colored paper construct a chart comparing the geocentric theory to the heliocentric theory of the universe.
2. The closer a star is to an observer the brighter it appears to be. Illustrate this point with two flashlights, one dim and one bright. Position one child with the bright flashlight some distance away. Have another child stand nearby with the dim flashlight. The closer child's light will appear brighter even though he has the dimmer light.
3. Make a booklet using fluorescent stars (available in variety stores) and dark construction paper to diagram different constellations.
4. Find Mars, Jupiter, Venus, or Saturn in the night sky (an almanac will tell you when these planets are visible). Observe the "wanderings" of these planets for several nights. Write a summary of your observations.

Section II Astronomy

1. Make a bulletin board about great scientists and their contributions to astronomy.
2. Have a contest with a classmate to name as many elliptical objects as you can.
3. Make a time line of famous astronomical discoveries.
4. With the astrolabe made in class, record the altitude of several prominent stars or the moon on several different clear nights.
5. Write a one-page report about one of these scientists: Aristotle, Ptolemy, Aristarchus, Copernicus, Brahe, Kepler, Newton, Jansky.

SECTION ONE

1.1 Any order:
   a. Most stars rise in the east, set in west.
   b. Sun and moon rise in east, set in west
   c. Sun rises and sets farther north in summer; farther south in winter.
   d. Moon does not always rise at the same time.

1.2 Any order:
   a. the sun and stars – day
   b. the motion of the moon – month
   c. the motion of sun against the stars – year

1.3 Either order:
   a. stars were guideposts for travelers
   b. stars helped man tell time

1.4 Example: "The heavens declare the glory of God; and the firmament sheweth his handiwork."

1.5 teacher check

1.6 position of stars changes

1.7 $186,000 \frac{mi.}{sec.} \times 60 \frac{sec.}{min.} \times 60 \frac{mi.}{hr.} \times 24 \frac{hr.}{day} \times 365 \frac{day}{year} = 5,865,696,000,000 \frac{mi.}{year}$

1.8 nine years from now

1.9 Light leaving the star now will take nine years to travel through space to earth.

1.10 teacher check

1.11 Sequence of answers may start anywhere.

Column 1:	Column 2:
The Fishes	The Ram
The Ram	The Bull
The Bull	The Twins
The Twins	The Crab
The Crab	The Lion
The Lion	The Maiden
The Maiden	The Scales
The Scales	The Scorpion
The Scorpion	The Archer
The Archer	The Goat
The Goat	The Water Carrier
The Water Carrier	The Fishes

1.12 teacher check

1.13 figure with head, arms, chest of a man and legs of a horse

1.14 a legend or story

1.15 lasting through the whole year

1.16 a pattern of stars

1.17 Any three:
Orion, Pleiades, Gemini, Taurus, Canis Major and Canis Minor

1.18 Altair, Vega, Deneb

1.19 all stars seen would be circumpolar

1.20 no stars seen would be circumpolar

1.21 Proxima Centauri, $4\ ^3/_{10}$ light years

1.22 sun

1.23 teacher check

1.24 teacher check

1.25 Both have earth at center, sun between Venus and Mars, fixed stars on one sphere and a Prime Mover. Ptolemy used epicycles to explain retrograde motion of planets. They made circles on circles with earth as center.

1.26
   a. when it is on the outside of your friend's circle
   b. when it passes between you and friend
   c. ball turning between you and friend appears to be moving backward as you watch friend
   d. when it passes directly between you and friend

1.27
   a. it would appear to move backward between earth and the line of its orbit
   b. at times they would be closer than at other times

1.28 student drawing like Figure 5

1.29 a. mass of rock or metal
      traveling through space

   b. mass of rock or metal that enters
       earth's atmosphere from space

   c. rock or metal that has
       reached earth from space

1.30 a. having to do with meteors

   b. having to do with meteorites

1.31 a. ° (circle 1 mm diameter)

   b. 100

1.32 metallic

1.33 (Optional activity)
      Teacher check

SECTION TWO

2.1 answers may vary

2.2 a. no

   b. they would have to
       travel together. One
       couldn't get ahead of
       or behind the other.

2.3 more complicated one

2.4 Circles within circles are more
     complicated than circles.

2.5 Complicated answers do not fit
     nature and are not usually accurate.

2.6 Any order:

   a. position of moon changes
       from night to night

   b. sun is eclipsed and
       moon is eclipsed

   c. planets do not move
       like stars do

2.7 We cannot feel the earth
     moving, but we can see the
     sun, moon, and stars "move."

2.8 Ptolemy requires epicycles
     to explain motions of
     planets. According to
     Copernicus, all the planets
     have the same motion.

2.9 Any order:

   a. Earth is not the center
       of the universe.

   b. Planets revolve around the sun and
       the sun is center of the solar system.

   c. Apparent motion of the
       sun is explained by the
       real motion of the earth.

2.10 Osiander's preface said
      Copernicus' cosmology was
      a method for calculating motions
      and not to be taken seriously.

2.11 Any order:

   a. earth wobbles

   b. route of earth around sun

   c. length of year to within 28 seconds

   d. diagrammed planets in
       correct order and determined
       approximate period of revolution

   e. recognized basic principle
       of relativity

2.12 teacher check

2.13 same size but clearer

2.14 the longer the tube the larger
      the image

2.15 to increase size of image

2.16 smaller inverted image

2.17 Galileo was convinced that you
      must observe *how* it happens before
      explaining what happens.

2.18 more interested in "what";
      not interested in making
      observations – just thinking

2.19 Galileo's refracting telescope
      used lenses; it was subject to
      blurring by aberration. Newton's
      reflecting telescope was shorter,
      but did not have the problem
      of blurring.

2.20 by placing a small mirror at 45°
      angle inside to reflect the
      image to the eyepiece

2.21 extremely careful observations over many years

2.22 Both believed the sun revolved around the earth.

2.23 He made very careful nightly observations that were later used in explaining the heavens.

2.24 The earth was the center (Ptolemy). All other planets revolved around sun (Copernicus).

2.25 He relied on Aristotle's theory that the orbits were perfect circles.

2.26 The ellipse becomes more circular.

2.27 no

2.28 having or representing the sun as the center

2.29 the point in a planet's orbit that is farthest from the sun

2.30 the point in a planet's orbit that is nearest the sun

2.31 teacher check

2.32 He built on the work of all the other great scientists.

2.33 The Greeks looked for the *why* of happenings. They depended on reasoning. Newton experimented and observed, trying to uncover the *how* of events.

2.34 Any order:
   a. every object pulls every other object
   b. gravitational pull depends on mass
   c. gravitational pull depends on distance between the two objects

2.35 Both the cow and the earth have mass and therefore attract other objects.

2.36 North Pole

2.37 Equator

2.38 North Pole

2.39 Equator

2.40 no

2.41 The part of the sky you see is defined by your horizon. As your horizon changes, the part of the sky you see changes.

2.42 a. Sun would cross zenith at noon.
   b. Sun would be halfway between highest and lowest altitude in sky at noon.
   c. Sun would be just rising to begin a six-month long day.

2.43 to measure altitude accurately

2.44 true north is the reference point to measure azimuth accurately

2.45 answers will vary

2.46 teacher check

2.47 astrolabe not level, astrolabe not lined to true north, inaccurate reading

2.48 INSTRUMENT, DATE OF INVENTION, NATION
   refracting telescope, 1608–1609, Holland – Italy
   reflecting telescope, 1669, England
   astrolabe, 150 B.C., Greece
   camera, 11th century, Europe
   radio telescope, 1931, United States
   spectroscope, 1860, Germany

2.49 to bend, as light passing through a lens

2.50 energy that is transmitted in the form of light, X rays, radio waves, and so on

SELF TEST 1

1.01 four

1.02 equator

1.03 earth

1.04 Aristotle

1.05 planets

1.06 wanderers

1.07 meteoroids

1.08 meteors

1.09 meteorites

1.010 Any order:

a. day; the motion of the turning earth

b. month; the motion of the moon

c. year; the motion of sun against the stars

1.011 Any order:

a. sun and moon across the sky

b. stars rise and set

c. sun's seasonal north-south progression

1.012 Any order:

a. rotation on its axis

b. revolution around the sun

c. solar system rotating on its axis

(or) galaxy moving away from other galaxies

1.013 Illustration 6

1.014 Illustration 7

1.015 a

1.016 e

1.017 h

1.018 c

1.019 l

1.020 d

1.021 f

1.022 i

1.023 b

1.024 j

1.025 k

1.026 They did not move as if they were fastened to spheres.

1.027 Any order:

a. Vega

b. Deneb

c. Altair

SELF TEST 2

2.01 Instrument: reflecting telescope
Operation: Light from moon reflects from large mirror, is bent by small 45° mirror, is focused at focal point, and is magnified by eyepiece.

2.02 Instrument: refracting telescope
Operation: Light from distant object comes through lens, is focused at focal point of objective lens, is magnified by eyepiece. Image is upside down.

2.03 Instrument: radio telescope
Operation: Radio waves from space are gathered and reflected toward focal point where receiver receives; waves are transmitted to recording device.

2.04 size

2.05 40

2.06 Yerkes

2.07 light years

2.08 $186{,}000 \frac{\text{mi.}}{\text{sec.}}$

2.09 6 trillion

2.010 radio

2.011 cameras

2.012 spectroscope

2.013 closer

2.014 Either order:
a. moons of planets
b. Earth's moon

2.015 6

2.016 a mirror

2.017 d

2.018 a

2.019 b

2.020 d

2.021 b

2.022 b

2.023 b

2.024 b

2.025 c

2.026 j

2.027 i

2.028 n

2.029 l

2.030 a

2.031 d

2.032 g

2.033 k

2.034 e

2.035 h

2.036 teacher check
The sun is at one focus. An ellipse describes the earth's path around sun.

2.037 When oriented to true north, full circle measures azimuth; half circle measures altitude above horizon. Gives location of star.

2.038 The approach to mysteries changed from one of argument and reason to one of experiment and observation.

2.039 Polaris would be beneath your feet, with 8,000 miles of earth in the way.

2.040 From the equator the horizon stretches from North Pole to South Pole. All stars appear to rise and set, so sometime during the year you could see each star.

SCIENCE 703
LIFEPAC TEST

1. a. orbits
    b. elliptical
2. rotation
3. day
4. Aristotle
5. sun
6. refracting
7. light
8. Galileo
9. light year (186,000 mi./sec.)
10. b
11. d
12. d
13. It helped man discover the nature of the heavenly bodies and their motions.
14. The larger the lens, the greater the magnification.
15. Because distances are so great it would be nearly impossible to work with such large numbers.
16. Aristotle did not experiment; he was interested in the *why*. Newton found experiments necessary to define the *what* and *how*.

17. a. First to use the telescope; "father of modern science"; found evidence for heliocentric theory
    b. First in modern times to promote the heliocentric theory
    c. Wrote laws of planetary motion
    d. Published the law of universal gravitation
    e. First to suggest that earth orbited sun
18. Any order:
    a. motion of sun and moon across sky
    b. motion against background of fixed stars
    c. seasonal motion north and south from horizon
19. geocentric: earth at center; heavenly bodies orbit earth. heliocentric; sun at center; bodies orbit sun.

Name _____

Match these items (each answer, 2 points).

1. _____ Galileo
2. _____ Cygnus
3. _____ meteoroid
4. _____ zenith
5. _____ Orion
6. _____ Ursa Major
7. _____ ellipse
8. _____ Copernicus
9. _____ Aristotle

a. heliocentric theory
b. Big Dipper
c. mass of rock in space
d. the Harp
e. Northern Cross
f. opposite of nadir
g. Betelguese
h. geocentric theory
i. orbit of planet
j. "Father of Modern Science"

Complete these statements (each answer, 3 points).

10. The motion of a planet around the sun is called a(n) _____ .
11. Telescopes that use mirrors for focusing light are called _____ telescopes.
12. The oval orbits traced by planets as they travel around the sun are_____ .
13. An astronomical instrument used for measuring the altitude of the sun and stars is a(n) _____ .
14. The height of a star, planet, or other heavenly body in the sky is its_____ .
15. A mass of rock or metal that enters the earth's atmosphere is a(n) _____ .
16. Elements in the stars can be identified by means of the _____ .

Answer these questions (each numbered item, 5 points).

17. What are Kepler's three Laws of Planetary Motion?
    a. _____
    b. _____
    c. _____
18. Why are the motions that we observe in the heavens only apparent motions?

_____
_____
_____

Complete these activities (each numbered item, 5 points).

19. Draw and label diagrams of the geocentric theory and the heliocentric theory.

20. Write an illustration of how the laws of gravitation and motion explain the motions of heavenly bodies. _____
_____
_____

21. List three flaws in the transparent sphere hypothesis.

 a. _____

 b. _____

 c. _____

22. List two reasons why the "wanderers" created problems in the geocentric theory.

 a. _____

 b. _____

55 / 69

Date _____

Score _____

1.   j
2.   e
3.   c
4.   f
5.   g
6.   b
7.   i
8.   a
9.   h
10. revolution
11. reflecting
12. ellipses
13. astrolabe
14. altitude
15. meteor
16. spectroscope
17. Any order:
   a. planets travel in elliptical orbits
   b. the speed of a planet moving around the sun changes all the time
   c. the period of a planet's revolution is related to the planet's distance from the sun
18. The earth is in action, but the earth's rotation and revolution seem to make the heavenly bodies move as we observe them.
19. See the drawings at the right.
20. Example:
   a cow tied to a post, the cow is the earth, the post is the sun, and the rope is gravity
21. Any order:
   a. position of moon changes from night to night
   b. sun is eclipsed and moon is eclipsed
   c. planets do not move like stars do
22. Either order:
   a. wanderers do not always move eastward against the background of the stars
   b. sometimes they seem to be closer, larger, and brighter than at other times

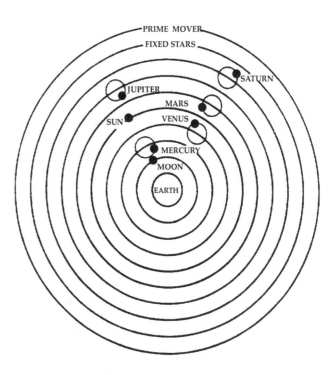

geocentric theory

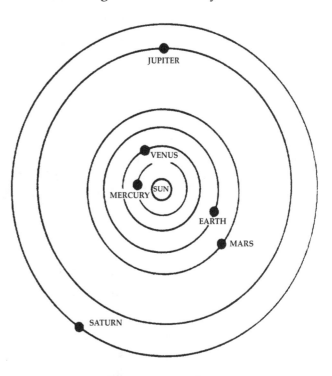

heliocentric theory

Materials Needed for LIFEPAC

Required:
package-sealing tape or reinforced
strapping tape
scissors
construction paper or banner paper

Suggested:
bicycle wheel
chalk
circle of corrugated cardboard,
30 cm in diameter
banner paper or several sheets of
construction paper taped together
string
protractor
red and black pencils
banner paper or any large sheet
of white paper
compass

ADDITIONAL LEARNING ACTIVITIES

Section I The Sun's Energy

1. Discuss the possible results if the sun were to explode and lose some of its mass.
2. Demonstrate the relationship between the distance from the sun and the amount of solar radiation received. You will need a 100-watt light bulb, a socket, a light meter (borrowed from a photographer), a meter stick, and graph paper. See Figure 1.

3. Construct a display of energy-saving devices. This might include items that require human energy rather than manufactured energy.
4. Using an encyclopedia or some other reference book study about one of these scientists and write a one-page report about his life and work: Hermann von Helmholtz, Albert Einstein, Hans A. Bethe.
5. Make a poster depicting the benefits of the sun's energy.
6. Make a graph representing your idea of how the amount of energy released by the sun changes with the passing of time.

Section II The Sun's Family

1. Create a mural depicting the planets, asteroids, and comets in the solar system.
2. Consult an almanac to learn when the next meteor shower will occur. Watch with a group of friends so a large area of the sky may be observed. Record the number of meteors you see.
3. Use a telescope to observe the planets.
4. Consult an almanac to learn when the planets will be visible and observe their change of position during several nights.
5. Use an encyclopedia to learn the sizes of the other moons in our solar system. Prepare a chart that compares our moon's size to the others.
6. Write a report about the rings of Saturn. What is their composition? Do other planets have a similar set of rings?

Section III The Moon

1. Construct a model with wire and styrofoam balls to show the inclination of the moon's orbit to the plane of the earth's orbit.
2. Make a diagram showing where low and high tides occur in relation to the moon's position.
3. Collect photographs of the moon and display them on a poster or in a scrapbook.
4. Observe and record the time of moonrise and moonset during the various phases of the moon.

Section IV Eclipses

1. Demonstrate an eclipse with a globe, a ball, and an electric light.
2. Learn that a circular object casts a round shadow with a ball, a square object, and a projector or flashlight. How would this investigation relate to the theory that the earth is round?
3. Make a chart to show the difference between a partial and a total eclipse.
4. Use an encyclopedia or almanac to learn how often eclipses occur and prepare a table showing this information.

Science 704 Answer Key

SECTION ONE

1.1 Any order:
  a. heat energy
  b. light energy
  c. photosynthesis
  d. comfortable temperature
    for man

1.2 teacher check

1.3 teacher check

1.4 false

1.5 false

1.6 true

1.7 true

1.8 false

1.9 a

1.10 a

1.11 c

1.12 Any order:
  a. Combustion – hydrogen and
    oxygen on the sun combine
    to give off energy
  b. Contraction – gravitation com-
    presses gases to generate
    energy
  c. Meteor – meteors falling into
    sun emitted energy
  d. Radioactivity – radioactive
    substances on the sun emit
    energy
  e. Carbon cycle – four hydrogen
    fuse to form one helium in
    the presence of carbon acting
    as a catalyst or
    Nuclear fusion – energy is
    released when hydrogen nuclei
    combine to form helium nuclei

1.13 by splitting the atom

1.14 The sun's energy is created by
  the nuclear fusion of hydrogen
  nuclei to form helium nuclei

1.15 one-two millionth

1.16 70 percent

SECTION TWO

2.1 Mercury
2.1 Mercury
2.3 fifty-nine days
2.4 eighty-eight days
2.5 the gravity
2.6 0.38 or slightly over ⅓
2.7 0.05 or ¹⁄₂₀
2.8 data table
2.9 Either order:
  a. Mercury
  b. Earth
2.10 108,230,000
2.11 phases
2.12 243 days

2.13 225 days
2.14 0.88
2.15 0.82
2.16 clouds do not change shape
2.17 data table
2.18 false
2.19 false
2.20 true
2.21 true
2.22 true
2.23 data table
2.24 data table
2.25 red color

40

2.26 228,000,000 km

2.27 687

2.28 24 hours 37 minutes

2.29 Either order:
   a. Phobos
   b. Deimos

2.30 0.39

2.31 0.11

2.32 revealed meteor craters; photo-graphed both moons, surface details, and a dust storm

2.33 a. no
   b. highs are never far above freezing

2.34 receives less than half heat and light, smaller than earth, less gravity, landscape is cratered

2.35 magnetic

2.36 Dark areas that appeared to be oceans; changes in landscape; apparent canals.

2.37 Hint:
   include information about temperature, atmosphere, surface gravity, distance (time to get there)

2.38 largest

2.39 778,400,000

2.40 318.3

2.41 2.64

2.42 9 hours 55 minutes

2.43 twelve years

2.44 a. four
   b. sixty-seven

2.45 red spot

2.46 Example:
   Jupiter's red spot is three times the diameter of the earth. The red spot was first seen in 1875 and has become fainter every year.

2.47 Hints for information to include in paragraph: measured radiation belts, reported amounts of hydrogen and helium, photographs of polar regions, new data on red spot, magnetic field, temperature

2.48 data table

2.49 1,424,600,000 km

2.50 a little more than ten hours

2.51 29½ years

2.52 1.17

2.53 95.3

2.54 its rings

2.55 twenty-three

2.56 Saturn's rings consist of thousands of narrow ringlets. These ringlets are made up of billions of ice particles that vary in size. The rings surround the equator of the planet. The rings are thin, measuring only 16 km in thickness.

2.57 Saturn is too far away. The trip would take at least five years. It is too difficult to equip a spaceship with power for that period of time. It would be too far from the sun to use solar energy.

2.58 data table

2.59 2,866,900,000 km

2.60 14.7

2.61 0.92

2.62 Sir William Herschel

2.63 North Pole faces sun for twenty years; then South Pole faces sun for twenty years. Rotation is clockwise.

2.64 Orbit around sun was not as expected; Astronomers deduced that another undiscovered planet must be pulling on it.

2.65 data table

2.66 4,486,100,000 km
2.67 about every 16 hours
2.68 165 years

2.69 Johann Galle

2.70 17.3
2.71 1.23

2.72 Neptune is slightly larger
      and colder – appears bluish;
      Uranus appears greenish
2.73 a. no
      b. Uranus still did not behave
          according to established
          laws. Still another planet
          had to be undiscovered.
2.74 data table
2.75 5,890,000,000
2.76 6 days 9 hours
2.77 248 years
2.78 the same as earth
2.79 0.15
2.80 Percival Lowell
2.81 it is so far away
2.82 data table
2.83 c
2.84 a
2.85 c
2.86 b
2.87 d
2.88 Asteroids (planetoids) are
      masses of rock orbiting the sun.
      They are between Mars and Jupiter.
      They shine with reflected sun-
      light and range in size from less
      than one km to 800 km.
2.89 Comets are made of frozen gases and
      dirt. They travel in elongated elliptical

orbits around the sun. Gases
form the tail, which is always
pushed away from the sun. They con-
tinuously lose some of their material
and will vanish, leaving particles
of dust that may enter the earth's
atmosphere as meteors.

2.90 1.2 cm

2.91 1.3 cm
2.92 0.7 cm
2.93 14.3 cm

2.94 12.0 cm

2.95 5.1 cm
2.96 4.9 cm
2.97 0.3 cm
2.98 Multiply your weight by the
      surface gravity factor for
      each celestial body.
2.99   0.38, teacher check
2.100 0.39, teacher check
2.101 2.64, teacher check
2.102 1.17, teacher check
2.103 0.92, teacher check
2.104 1.23, teacher check
2.105 0.15, teacher check
2.106 0.16, teacher check
2.107 a. 57,900,000
       b. 4,878
       c. 88 days
       d. 59 days
       e. 0.05
       f. 0.38
       g. 0

2.108 a. 108,230,000
         b. 12,100
         c. 225 days
         d. 243 days
         e. 0.82
         f. 0.88
         g. 0

2.109 a. 150,000,000
         b. 13,000
         c. 365.25 days
         d. 23 hours 56 minutes
         e. 1.00
         f. 1.00
         g. 1

2.110 a. 228,000,000
         b. 6,760
         c. 687 days
         d. 24 hours 37 minutes
         e. 0.11
         f. 0.39
         g. 2

2.111 a. 778,400,000
         b. 142,700
         c. 12 years
         d. 9 hours 55 minutes
         e. 318.3
         f. 2.64
         g. 67

2.112 a. 1,424,600,000
         b. 120,000
         c. 29 ½ years
         d. 10 hours

         e. 95.3
         f. 1.17
         g 23

2.113 a. 2,866,900,000
         b. 50,800
         c. 84 years
         d. 17 hours
         e. 14.7
         f. 0.92
         g. 15

2.114 a. 4,486,100,000
         b. 48,600
         c. 165 years
         d. about 16 hours
         e. 17.3
         f. 1.23
         g. 1

2.115 a. 5,890,000,000
         b. 3,000
         c. 248 years
         d. 6 days 9 hours
         e. 1.0
         f. 0.15
         g. 1

2.116 a. —
         b. 3,476
         c. —
         d. 27 ⅓ days
         e. —
         f. 0.16
         g. —

SECTION THREE

3.1 circular
3.2 circular
3.3 circular
3.4 Example: looks like a series of half circles up
3.5
3.6 epicycles
3.7 yes
3.8
3.9 a. the earth
      b. moon
      c. sun

3.10 Viewed from the earth, the moon moves in a circle. Viewed from the sun, the moon moves in epicycles.
3.11 The line describes one-quarter of a wavy ellipse around the "sun."
3.12 false
3.13 false
3.14 true
3.15 true
3.16 false
3.17 a. east
      b. west

3.18 of the earth's rotation and the moon's revolution

3.19 thirteen

3.20 13° or about 50 minutes of time

3.21 50 minutes

3.22 27 ⅓ days

3.23 29 ½ daysand

3.24 because while the moon is revolving around the earth, the earth is revolving around the sun at the rate of 1° per day.

3.25 new moon to new moon (29 ½ days)

3.26 daytime

3.27 a. no
b. the lighted side would always be facing the sun – the moon receives its light from the sun.

3.28 The tides are the periodic rise and fall of the level of the ocean.

3.29 gravitational pull of the moon and the sun

3.30 moon

3.31 The same point on earth is directly beneath the moon every 24 hours and 52 minutes. A high tide occurs at that point (nearly) and on the opposite side of the earth.

3.32 moon and sun are in a line and pulling together

3.33 a. tides during full and new moon – highest highs and lowest lows
b. tides during first third quarters – low high tides and high low tides

3.34 difference between level of water at low tide and level of water at high tide

3.35 one-half of earth would always be day – other half always night

3.36 seasons would not change

3.37 method of telling time by motions in heavens would cease

3.38 lighted side would get very hot – dark side very cold;

3.39 would be pulled to one side of the earth

3.40 could not exist as we know it today

3.41 earth would fall into the sun

SECTION FOUR

4.1 Because the sun, even though larger, is so far away it *appears* to be the same size as the moon.

4.2 a. 384,000 km ÷ 3,476 km = 110.47
b. 152,000,000 km ÷ 1,392,000 km = 109.19
c. If they have the same angular diameters, they must have the same ratio of distance and diameter.

4.3 teacher check

4.4 6 ½ cm

4.5 no

4.6 no

4.7 a. no
b. the moon's shadow is cast on only a small portion of the earth's surface

4.8 The moon is far enough away from the earth that the earth's curvature does not block the line of sight

4.9 The earth must be between the sun and the moon.

4.10 The moon must be between the sun and the earth.

4.11 a. the moon gradually passes into earth's shadow (lunar eclipse)
b. an eclipse of the sun

4.12 See Figure 14 in text.

SELF TEST 1

1.01   d

1.02   f

1.03   a

1.04   i

1.05   h

1.06   c

1.07   g

1.08   e

1.09   weather

1.010   coal

1.011   photosynthesis

1.012   oxygen

1.013   nuclear

1.014   Einstein

1.015   c

1.016   a

1.017   b

1.018   f

1.019   e

1.020   d

1.021   Example:
   a. proper temperature
   b. provides photosynthesis
   c. stored energy as coal, petroleum
   d. causes evaporation and precipitation
   e. heats some homes directly

SELF TEST 2

2.01   b

2.02   c

2.03   c

2.04   b or c

2.05   c

2.06   a

2.07   a

2.08   b

2.09   nine

2.010   counterclockwise

2.011   Pluto

2.012   Jupiter

2.013   Jovian (outer)

2.014   Venus

2.015   150,000,000

2.016   Saturn

2.017   away from

2.018   wanderers or *planetae*

2.019   rotation

2.020   day

2.021   elliptical

2.022   terrestrial or inner

2.023   sun

2.024   Photosynthesis is necessary for plants to produce food. This food is necessary to animal life either directly through eating plants or indirectly through eating animals that eat plants. Photosynthesis depends on light energy. We could not live without heat energy to warm Earth, create climate, and so forth.

2.025   A six-step reaction in which four hydrogen nuclei fuse to form one helium nucleus. The carbon atom is involved as a catalyst. Vast energy is released in the carbon cycle.

2.026   They are on the sun side of the earth.

2.027   They are on the daytime side of the earth. They are almost in a direct line between earth and sun and cannot be seen so close to the sun.

SELF TEST 3

3.01 kilometers

3.02 counterclockwise

3.03 Jovian or outer

3.04 Saturn

3.05 away from the sun

3.06 29 ½

3.07 27 ⅓

3.08 29 ½

3.09 a. moon
Either order:
b. sun
c. earth

3.010 wanderers or *plantae*

3.011 a. rotates
b. revolution

3.012 sun

3.013 6 hours 13 minutes

3.014 rotation

3.015 b

3.016 d

3.017 i

3.018 e

3.019 j

3.020 c

3.021 f

3.022 g

3.023 h

3.024 a

3.025 While the moon revolves around the earth for 27 ⅓ days, the earth has moved along part of its curved path around the sun. The additional two-day journey of the moon brings it between the sun and the earth again. (See illustration on page 37.)

3.026 A wavy ellipse, with the sun at one focus of the ellipse.

SELF TEST 4

4.01 ½

4.02 full

4.03 new

4.04 lunar

4.05 solar

4.06 corona

4.07 hydrogen

4.08 Mercury

4.09 gravitation

4.010 counterclockwise

4.011 full

4.012 6 hours 13 minutes

4.013 elliptical

4.014

4.015

4.016 The moon casts a small shadow.

4.017 b

4.018 a

4.019 c

4.020 d

4.021 c

SCIENCE 704
LIFEPAC TEST

1. g
2. c
3. i
4. a
5. l
6. e
7. h
8. n
9. d
10. b
11. p
12. o
13. k
14. q
15 m
16. j
17. f
18. coal
19. photosynthesis
20. fusion
21. temperature
22. weather
23. gravity
24. Either order:
    a. Mercury
    b. Earth
25. phases
26. 150,000,000 km
27. axis
28. Mars
29. Galileo
30. Saturn
31. Uranus
32. Neptune
33. a. Venus
    b. Neptune
34. Pluto
35. 1986
36. Jupiter
37. Either order:
    a. Mercury
    b. Venus

38. counterclockwise
39. east
40. 13°
41. later
42. a. 27 $\frac{1}{3}$
    b. 29 $\frac{1}{2}$
43. 29 $\frac{1}{2}$ days
44. a. Moon
    Either order:
    b. Earth
    c. Sun
45. straight line
46. solar eclipse
47.
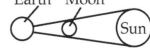
48.
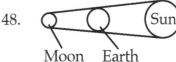
49. Examples:
    a. light
    b. photosynthesis
    c. heat
    d. fuel or weather
    or electricity and so forth
50. Any order: any five
    a. combustion
    b. contraction
    c. meteor impact
    d. radioactive substances
    e. carbon cycle
    f. proton-proton
51. a. Mercury
    b. Venus
    c. Earth
    d. Mars
    e. Jupiter
    f. Saturn
    g. Uranus
    h. Neptune
    i. Pluto

Name _____

Answer *true* or *false* (each answer, 1 point).
1. _____ The tail of a comet always points toward the sun.
2. _____ The energy that turns the blades of a windmill is derived from solar energy.
3. _____ The 29 ½-day interval from new moon to new moon is called a lunar month.
4. _____ Asteroids orbit between Jupiter and Saturn.
5. _____ The splitting of atoms that releases tremendous amounts of energy is fission.
6. _____ Nitrogen is the most abundant gas in the earth's atmosphere.
7. _____ Tides are caused by the gravitational pull of the earth.
8. _____ The point closest to the sun in the orbit of a planet is aphelion.
9. _____ Petroleum is a form of stored solar energy.
10. _____ A new moon occurs when the earth is between the sun and the moon.

Match these items (each answer, 2 points).
11. _____ Jupiter
12. _____ Mars
13. _____ corona
14. _____ fusion
15. _____ spring tides
16. _____ combustion
17. _____ neap tides
18. _____ photosynthesis
19. _____ Venus
20. _____ nucleus

a. releases oxygen
b. low tides
c. "red planet"
d. Morning Star
e. penumbra
f. fifth planet from sun
g. center of atom
h. process of burning
i. high tides
j. joining helium and hydrogen nuclei
k. halo around the sun

Complete these statements (each answer, 3 points).
21. When the moon blocks the sun's light a(n) _____ occurs.
22. In our solar system are two sets of sister planets. One set is Earth and a. _____ , and the other set is Uranus and b. _____ .
23. The two planets that show phases like the moon are a. _____ and b. _____ .
24. Earth is located between a. _____ and b. _____ .
25. The moon rises and sets fifty minutes _____ each day.
26. The sun's energy is the result of _____ .

27. The position of the sun in the sky determines the _____ of a location.
28. The moon appears to rise in the _____ .
29. We see only one side of the moon because it a. _____ only once while it is making one complete b. _____ .
30. Our solar system consists of _____ planets.

Complete these activities (each numbered item, 5 points).

31. List the planets of our solar system in order from the sun.

    a. _____   f. _____
    b. _____   g. _____
    c. _____   h. _____
    d. _____   i. _____
    e. _____

32. List five theories of how the sun produces solar energy.

    a. _____
    b. _____
    c. _____
    d. _____
    e. _____

33. Make a drawing of a lunar eclipse.

34. Make a drawing of a solar eclipse.

74 / 92

Date _____

Score _____

1. false
2. true
3. true
4. false
5. true
6. true
7. false
8. false
9. true
10. false
11. f
12. c
13. k
14. j
15. i
16. h
17. b
18. a
19. d
20. g
21. eclipse or solar eclipse
22. a. Venus
    b. Neptune
23. Either order:
    a. Mercury
    b. Venus

24. Either order:
    a. Venus
    b. Mars
25. later
26. nuclear reaction, fusion or carbon cycle
27. temperature
28. east
29. a. rotates
    b. revolution
30. nine
31. a. Mercury
    b. Venus
    c. Earth
    d. Mars
    e. Jupiter
    f. Saturn
    g. Uranus
    h. Neptune
    i. Pluto
32. Any order, any five:
    a. combustion
    b. contraction
    c. meteor-impact
    d. radioactive substances
    e. carbon cycle
    f. proton-proton

33.

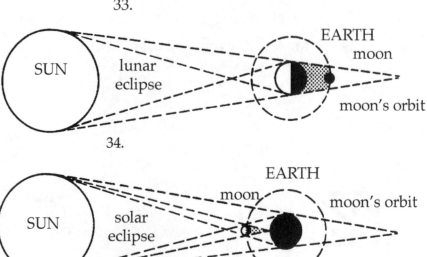

34.

Materials Needed for LIFEPAC

Required: Suggested:

None   None

Additional Learning Activities

Section I The Universe

1. Work numerous astronomical distance problems incorporating scientific notation to ensure student comprehension of superscript computations when large numbers are multiplied and divided.
2. Have the class make mathematical comparisons among the planets Mars, Venus, and Earth with respect to their mass, density, temperature, and surface gravity.
3. With a friend identify the constellations at night. Sketch and label the different constellations.
4. With a friend visit a local planetarium.
5. Obtain a picture or schematic representation of the Milky Way Galaxy which shows where our solar system fits into its vast elliptical dimensions.
6. Research the Greek astronomers. Describe the difference between ancient Greek and present-day astronomy.

Section II Telescopes and Optics

1. Show the class refracting and reflecting telescopes and explain the difference. If a physics laboratory is available which has a prism experiment, refraction can be more readily explained than through the use of a diagram.
2. With a friend look through a telescope and chart a constellation.
3. Research the life of Isaac Newton. Report to the class on his importance to the science of astronomy.
4. Research the critical new developments in telescopes. Write a one-page paper on the new developments.

Section III Space Exploration

1. Lead a class discussion concerning the estimate of statistical chance for the existence of extraterrestrial intelligent life.
2. Use the delphi-technique to attempt to get student input into technology breakthroughs required for interstellar travel.
3. Write a one-page report on what kind of life might exist on the other planets.

SECTION ONE

1.1 false

1.2 true

1.3 false

1.4 a

1.5 b

1.6 c

1.7 For six of the known planets, Bode's Law is a very close approximation in A. U. from the actual mean distance since confirmed.

1.8 An astronomical unit is the distance of the earth from the sun, or 93,000,000 miles.

1.9 count the number of zeros in a number and write it as a superscript to the base number 10.

1.10 teacher check

1.11 false

1.12 true

1.13 false

1.14 c

1.15 d

1.16 a

1.17 A linear increase in brightness (1 mag. diff.) according to the eye is precisely measured as a geometric increase in brightness (2.51 times brighter).

1.18 sixth magnitude

1.19 The parallax technique is used to indirectly measure the distance to far away objects. The distance to the star is relative to its parallax angle, which is one-half the apparent change in its angular position during a six-month period.

1.20 a. Carina
b. Camelopardalis
c. Cassiopeia

1.21 because one is dealing in relative distances

1.22 If the difference in apparent magnitude (2.3) is rounded to 2, the answer is 6.3. There are two ways to solve the problem with more precision.

a. Plot the data on top of page 7 and then plot the magnitude difference (-1.5-(+0.8) or 2.3 on the graph. The brightness ratio will equal about 8.3.

b. The second way is to use the method shown on page 7.
$$2.51^{2.3} = (2.51)^{0.1} \times (2.51)^{0.1} \times (2.51)^{0.1} \times (2.51)^{2.0}$$
$$= 1.1 \times 1.1 \times 1.1 \times 6.3 = 8.3$$

1.23 Draco is farther away by about 1.81 LY.

Solution:

Alpha Draco: $pc = \dfrac{1}{parallax} = \dfrac{1}{0.18''}$

$= 5.56\ pc$

$5.56\ pc \times 3.26\ \dfrac{LY}{pc} = \underline{18.13\ LY}$

Altair: $pc = \dfrac{1}{parallax} = \dfrac{1}{20''}$

$= 5.00\ pc$

$5.00\ pc \times 3.26\ \dfrac{LY}{pc} = \underline{16.30\ LY}$

Distance difference
$= 18.13\ LY - 16.30\ LY$
$= \underline{1.83\ LY}$

1.24 6×10^{13} miles

SECTION TWO

2.1   aperture
2.2   one-third inch
2.3   Area of 2″ lens $= \pi\left(\dfrac{d}{2}\right)^2$

$$= \pi\left(\dfrac{2}{2}\right)^2$$

$$= \dfrac{\pi}{1}$$

Ratio of areas $= \dfrac{\pi}{36} \div \dfrac{\pi}{1}$

$$= \dfrac{\pi}{36} \quad \left(\dfrac{1}{\pi}\right)$$

$$= 1{:}36$$

A 2-inch telescope is 36 times as powerful as the unaided eye.

2.4   d
2.5   a
2.6   f
2.7   e
2.8   h

2.9   i
2.10   chromatic aberration
2.11   flint
2.12   30X
2.13   see figure 5
2.14   see figure 6
2.15   spectrograph
2.16   Either order:
   a.   strength of hydrogen line (chemical composition)
   b.   surface temperature
2.17   O
2.18   Q
2.19   Arecibo
2.20   relatively small celestial object which shines brighter than a hundred normal galaxies

SECTION THREE

3.1   c
3.2   d
3.3   g
3.4   a
3.5   b
3.6   e
3.7   Either order:
   a.   force of gravity
   b.   centrifugal force
3.8   second
3.9   Either order:
   a.   the velocity of the earth's rotation about its axis
   b.   its best range of latitudes on the earth's surface for satellite observation
3.10   Any order:
   a.   communications satellite
   b.   weather satellite
   c.   earth-resources satellite
   d.   military satellite

3.11   to map the existence of certain mineral deposits and the extent of other natural resources
3.12   Evidence that a great deal of water once flowed on Mars. Mars is considered to be in the sun's habitable zone. Some forms of earth life can survive Martian environmental conditions.
3.13   Some form of activity was present in soil samples, but scientists cannot explain whether it is because of living organisms or some very unusual chemical characteristics.
3.14   that the natural laws that govern earth's physical and chemical events apply universally

SELF TEST 1

1.01 d
1.02 a
1.03 f
1.04 b
1.05 g
1.06 Either order:
    a. Mercury
    b. Mars
    c. Venus
1.07 Any order:
    a. Mercury
    b. Mars
    c. Earth
    d. Jupiter
    e. Venus

    f. Saturn
1.08 18 trillion
1.09 3.3
1.010 3 millionths
1.011 2.5:1
1.012 1/3600
1.013 2
1.014 a. Cassiopeia
    b. Big Dipper (Ursa Major)
    c. Draco
    d. Cepheus
    e. Ursa Minor
    f. Ursa Major
    g. Camelopardalis
    h. Lynx

SELF TEST 2

2.01 false
2.02 true
2.03 true
2.04 false
2.05 true
2.06 d
2.07 a
2.08 f
2.09 c
2.010 b
2.011 g

2.012 i
2.013 2,500 times brighter or 1:2,500
2.014 660 inches
2.015 2400X
2.016 a. one-tenth
    b. 400
2.017 19
2.018 Pluto
2.019 Southern Hemisphere
2.020 light-year

SELF TEST 3

3.01 false
3.02 true
3.03 false
3.04 false
3.05 true
3.06 e
3.07 a
3.08 g
3.09 f
3.010 h
3.011 b

3.012 c
3.013 d
3.014 the basic building blocks of life
3.015 Either order:
    a. physiological
    b. biological
3.016 a. 1400
    b. 1730
3.017 2.8
3.018 Galileo

3.019 (1) raising it to the proper elevation above the earth's surface; (2) orienting it in the proper attitude; and (3) giving it the proper speed

3.020 the great circle on the celestial sphere that passes through the celestial poles and the observer's zenith

3.021 The astronomical unit is the distance of the earth from the sun or 93,000,000 miles.

3.022 Stars are large globes of intensely heated gas, which generate their own light. Nebulae are vast clouds of dust and gas made visible by the light of stars.

3.023 Skylab 4

3.024 Voyager

3.025 Pioneers 10 and 11

3.026 Sputnik

3.027 Apollo

Science 907
LIFEPAC Test

1. satellite

2. Either order:

   a. Venus

   b. Mercury

3. 1 million (1,000,000)

4. Alpha-Centauri

5. Neptune

6. parallax

7. sixth

8. refracting

9. blue

10. increase the focal length

11. focal length

12. achromatic refractors

13. silver

14. 236 inches

15. Pioneer

16. SETI

17. 100 years

18. d

19. b

20. a

21. c

22. f

23. g

24. e

25. 500 A.U. $= 5 \times 10^2$ A.U. $\times 93 \times 10^6 \dfrac{\text{mile}}{\text{A.U.}}$

$= 4.65 \times 10^{10}$ mile

1 LY $= 6 \times 10^{12}$ mile therefore,

500 A.U. $= \dfrac{4.65 \times 10^{10} \text{ mile}}{6 \times 10^{12} \text{ mile/LY}}$

$= 0.78 \times 10^{-2}$ LY

$= 7.8 \times 10^{-3}$ LY

26. c

27. f

28. a

29. e

30. b

Name _____

Complete these statements (each answer, 3 points).

1. When we employ the power of base 10 to write very large or very small numbers, we are using _____ .
2. The moons which orbit their planets are called _____ .
3. In increasing order of the mean distance from the sun, the major planets are

   a. _____ ,   b. _____ ,   c. _____ ,
   d. _____ ,   e. _____ ,   f. _____ ,
   g. _____ ,   and h. _____ .
4. The distance of 93,000,000 miles is called a(n)_____ .
5. A light-year measures_____ miles.
6. The star which has an apparent magnitude of +0.8 is _____ .
7. An optical telescope used to pick up the faintest visible stars is known as a _____ telescope.
8. The magnifying power of a telescope is computed by dividing the focal length of its objective by _____ .
9. In the Newtonian Reflector, light rays are bent at _____ before the image is formed.
10. Mars was explored by unmanned spacecraft called _____ .
11. Skylab 4 placed astronauts into weightlessness for a period of _____ days.
12. The radio telescope at _____ measure 1,000 feet along its dish diameter.
13. Pioneer 10 and 11 made several passes around the planet _____ .
14. When we speak of life possibly outside of Earth, we use the term

    _____ .
15. Apollo 11 made a _____ landing.
16. Only the _____ has successfully landed men on the moon.
17. In order to undertake interstellar travel, we would need spacecraft with speeds near _____ .

Match these items (each answer, 2 points).

18. _____ parallax
19. _____ recombines colors
20. _____ Chester Hall
21. _____ James Gregory
22. _____ Galileo's telescope
23. _____ 3-inch, f/8
24. _____ colored halo effect

a. theory behind achromatic refractors
b. suggested reflecting telescope
c. technique for measuring star distance
d. flint glass
e. 24-inch focal length
f. discovered gravity
g. 30X
h. chromatic aberration

Complete this activity (this answer, 5 points).

25.   Calculate the degree of magnification for a 2″ and 5″ lens.

Complete this table (this answer, 3 points).

Celestial Body	*Rotation Period (days)*
26. Earth	_____
27. Sun	_____
28. Mars	_____
29. Venus	_____

85 / 106

Date _____

Score _____

1.   scientific notation
2.   satellites
3.   a. Mercury
     b. Venus
     c. Earth
     d. Mars
     e. Jupiter
     f. Saturn
     g. Uranus
     h. Neptune
4.   Astronomical Unit
5.   6 trillion
6.   Betelgeuse
7.   reflecting
8.   the focal length of its eyepiece
9.   right angles
10. Viking I and II
11. 84 days
12. Arecibo
13. Jupiter
14. extraterrestrial
15. lunar
16. United States
17. the velocity of light
18. c
19. d
20. a
21. b
22. g
23. e

24. h
25. $\dfrac{\pi(2/2)^2}{\pi(5/2)^2} = \dfrac{1}{25/4} = \dfrac{4}{25}$

     Therefore, 4:25.
26. 1
27. 27
28. 1
29. -243